4/21/00
W9-AZG-512
WITHDRAWN

BELMONT PUBLIC LIBRARY

THE ELEMENTS

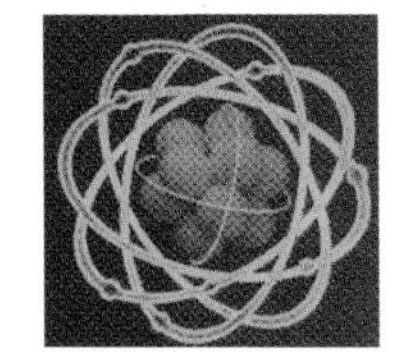

Hydrogen

John Farndon

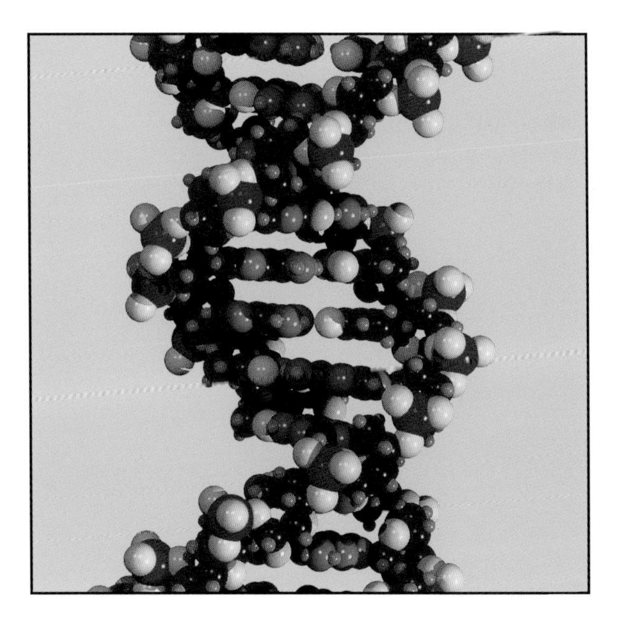

BENCHMARK BOOKS

MARSHALL CAVENDISH
NEW YORK

A6W-9342

J
546.2
FAR

Benchmark Books
Marshall Cavendish Corporation
99 White Plains Road
Tarrytown, New York 10591-9001

© Marshall Cavendish Corporation, 2000

All rights reserved. No part of this book may be reproduced or utilized in any form
or by any means electronic or mechanical including photocopying, recording,
or by any information storage and retrieval system, without permission from the copyright holders.

Library of Congress Cataloging-in-Publication Data
Farndon, John.
Hydrogen / by John Farndon.
p. cm. — (The elements)
Includes index.
Summary: Explores the history of the chemical element hydrogen and
explains its chemistry, how it reacts, its uses, and its importance in our lives.
ISBN 0-7614-0886-X (lib. bdg.)
1. Hydrogen—Juvenile literature. [1. Hydrogen.] I. Title.
II. Series: Elements (Benchmark Books)
QD181.H1F33 2000
546'.2—dc21 98-44692 CIP AC

Printed in Hong Kong

Picture credits
Corbis (UK) Ltd: 4, 8, 13, 20.
Image Bank: 5, 6, 9 *top*, 11, 19.
Science Photo Library: 7, 9 *bottom*, 12, 14, 15, 16, 17, 21, 22, 23, 24, 26, 30.

Series created by Brown Packaging Partworks
Designed by wda

Contents

What is hydrogen?

Hydrogen is an amazing substance. It is the lightest of all gases—an olympic swimming pool full of hydrogen would weigh barely 2 lbs. (1 kg). It has no taste, smell, or color. Its atom is the smallest and simplest of all atoms. Yet it is the most important substance in the Universe.

Almost all the matter in the Universe is hydrogen. It was the first element to appear, soon after the Universe began with the "Big Bang." It took billions of years for most other elements to form. Hydrogen atoms are the basic building blocks from which all other atoms are made.

On Earth, hydrogen is a component of life's most vital ingredient—water. Water molecules are made from two hydrogen atoms joined to one oxygen atom.

The hydrogen atom

Hydrogen is the first element in the periodic table, with an atomic number of one. This means that in the nucleus, at the center of each hydrogen atom, is a single, tiny particle called a proton. Protons have a positive electrical charge. A negatively charged particle, called an electron, circles the nucleus in an electron shell.

Most hydrogen atoms have an atomic mass of 1.00794. But one in every 6,000 or so hydrogen atoms also has a neutron in

Without water, life on Earth would be impossible. Each and every molecule of water contains two atoms of hydrogen.

DID YOU KNOW?

METAL OR NONMETAL?

Chemists cannot decide if hydrogen is a metal or a nonmetal. Because it has just one electron, it behaves in some ways like the metals called "alkali metals," which include lithium and sodium, and which all have a single electron in the outermost shell of their atoms. On the other hand, only two electrons will fit in the shell of a hydrogen atom. So in other ways, hydrogen behaves like a member of the halogens. This is a group of nonmetals, which includes the gases fluorine and chlorine, all of which have one electron missing. So depending on whether you look at the hydrogen atom's electron shell as half full or half empty, you can classify hydrogen either as a metal or a nonmetal.

its nucleus. These atoms are twice as heavy as normal hydrogen atoms. Hydrogen atoms with *two* neutrons in the nucleus can be made artificially. These atoms are three times heavier than the normal ones.

The Sun may look solid, but it is really a huge fireball of gases, chief among them hydrogen.

HYDROGEN ATOM

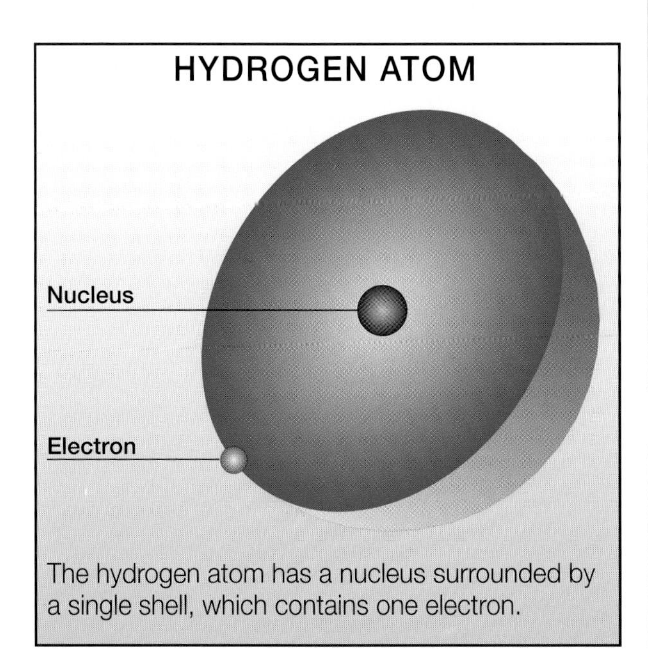

The hydrogen atom has a nucleus surrounded by a single shell, which contains one electron.

ATOMS AT WORK

The hydrogen nucleus is circled by a single electron. The atom needs two electrons to make it stable.

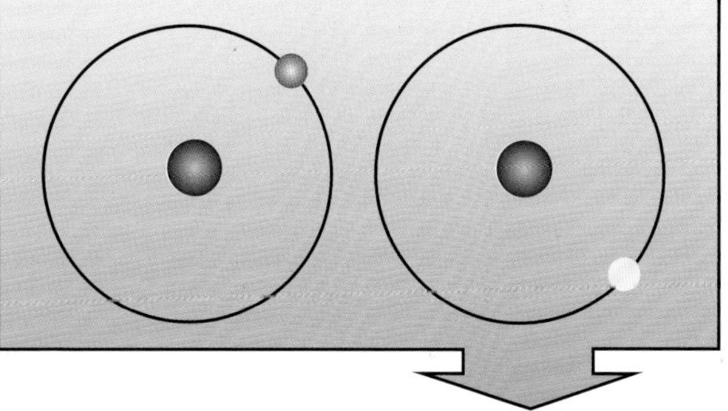

When one hydrogen atom meets another, they share their electrons with each other so that both have a full complement of two electrons.

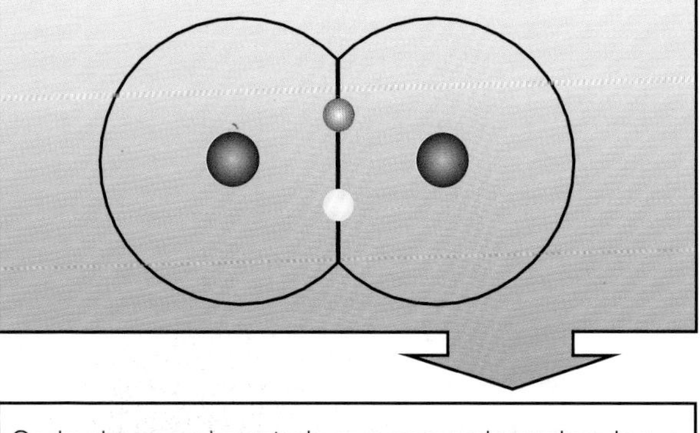

So hydrogen almost always occurs in molecules consisting of pairs of atoms, called "diatomic" molecules. Its formula is written like this:

H_2

Where hydrogen is found

Hydrogen is by far the most abundant element, making up almost 90 percent of all the atoms in the Universe. Most of the remaining 10 percent of all atoms are helium atoms, so helium is the only element remotely comparable in abundance. The Sun and every other star, the giant clouds of gas between stars, and most large planets are essentially made of hydrogen, with a little helium. Indeed, hydrogen is the nuclear fuel that makes stars glow. Deep in the heart of most stars, the nuclei of hydrogen atoms are squeezed together under intense pressure to make helium atoms. This process releases huge quantities of nuclear energy.

Hydrogen is not the dominant element on Earth simply because our planet is so small and hydrogen is so light. Most of the hydrogen that was once present on Earth has drifted off into space. Even so, hydrogen is the ninth most abundant element in Earth's crust, making up almost one percent of it by weight.

On Earth, hydrogen can be found in many compounds. The water around this oil rig is a compound with oxygen. The oil itself is a hydrocarbon—a compound of hydrogen and carbon.

When the Milky Way, seen above, came into being, the gas that filled the Universe was hydrogen.

Most of the hydrogen on Earth occurs in combination with other elements, notably with oxygen—as water. Its importance as one of the two elements in water cannot be overestimated, since water covers three-quarters of the world's surface and makes up three-quarters of our bodies.

ABUNDANCE IN PARTS PER MILLION	
The Universe	900,000
Sun	745,000
Streams, rivers, and lakes	111,900
Seawater	107,800
Human body	100,000
Earth's crust	1,400

Pure hydrogen gas exists naturally only in a few places, mainly in underground pockets. It does exist in the atmosphere, but only in minute quantities—less than one part in a million by volume.

Hydrogen compounds play a central role in all living things. Sugars, amino acids, proteins, cellulose, and many other biologically important molecules contain hydrogen. Together with carbon and oxygen, it is one of the basic elements of life. Fossil fuels, such as coal, gasoline, and natural gas, are also compounds of hydrogen and carbon (we term these compounds *hydrocarbons)*.

Special characteristics

In nearly all conditions on Earth, hydrogen is a gas. Under normal pressures, it does not condense (turn to liquid) until it is very cold. Its boiling point is -423.17°F (-252.87°C), the lowest of any substance. It only turns solid when the temperature drops to -434.45°F (-259.14°C), which is colder than just about anywhere in the Universe.

However, under extreme pressure, hydrogen will turn into a liquid and even a solid. Solid hydrogen has been made in laboratory experiments. It looks and behaves like a metal. Jupiter, the largest planet in our solar system, is made largely of hydrogen. As it is so huge, Jupiter's gravitational pull is enormous, creating extreme pressures at its surface. Beneath Jupiter's thin atmosphere, there is probably an ocean of liquid hydrogen some 15,500 miles (25,000 km) deep. At the bottom of this ocean, the pressure is some three million times greater than that on Earth's surface, and the hydrogen is squeezed so hard it becomes a metal.

The planet Jupiter, seen above, has an atmosphere of hydrogen gas, beneath which lies a deep ocean of liquid hydrogen.

DID YOU KNOW?

HYDROGEN METAL

In a groundbreaking experiment in 1972, scientists at Livermore in California briefly managed to create a dramatic change in hydrogen's density when they subjected it to a pressure two million times greater than that on Earth's surface. In 1973, a Russian team believed they had created metallic hydrogen at a pressure 2.8 million times that on Earth's surface. If hydrogen metal was created in these experiments, it existed only for a moment, and little was discovered about it. In 1996, however, scientists in Britain were able to make and study hydrogen metal, which is believed to be by far the best possible conductor of electricity.

How hydrogen was discovered

English scientist Robert Boyle was the first person to isolate the gas that was later named hydrogen.

English chemist Robert Boyle (1627–1691), the youngest of the 14 children of the first Earl of Cork, was one of the most brilliant scientists of all time. Called "the father of modern chemistry," he devised the first definition of the elements, as "primitive and simple, or perfectly unmingled bodies."

In 1671, Boyle wrote a paper titled "New experiments touching the relation betwixt flame and air." This paper was about the nature of burning. In it, Boyle described how iron filings react with dilute acids to create a gas. The gas burned so readily that he called it "inflammable solution of Mars" (Mars was his word for iron.) The gas was hydrogen.

In 1766, British chemist Henry Cavendish (1731–1810) realized that the gas was an element, which is why he is credited with its discovery. Cavendish made hydrogen by pouring acid onto mercury metal and collecting the gas that was released. He called the gas "inflammable air." It was French chemist Antoine Lavoisier (1743–1794) who named it hydrogen.

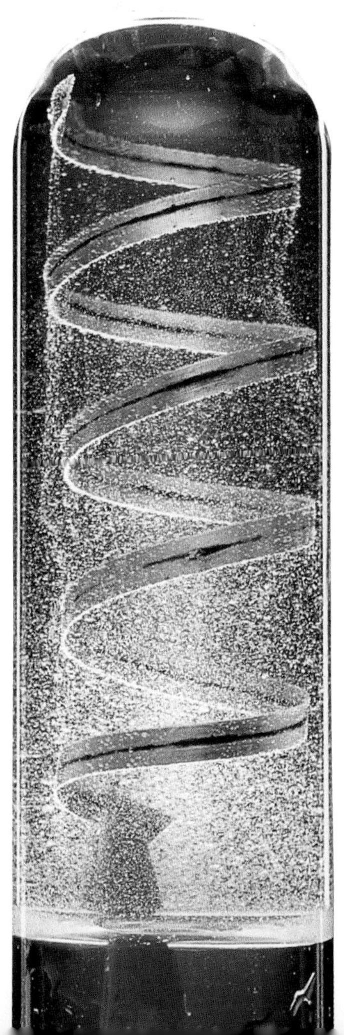

Inside this glass tube is a twisted ribbon of the metal magnesium. Visible bubbles of hydrogen gas are given off as the magnesium ribbon reacts when hydrochloric acid is added to the tube.

DID YOU KNOW?

ANTOINE LAVOISIER

French scientist Antoine Lavoisier, who gave hydrogen its name—from two Greek words meaning "water-making"—was guillotined in 1794, at the height of the French Revolution.

How hydrogen reacts

Hydrogen is one of the most reactive of all gases. It will readily react with nearly all other elements if it comes in contact with them in the right circumstances. Because hydrogen atoms are so small and light, they easily mix in with other atoms, so many of these reactions will happen with just a small quantity of hydrogen.

DID YOU KNOW?

AIRSHIPS

Hydrogen is the lightest of all substances, with a density of just 0.084 grams per liter. Such a light gas seemed perfect for lifting balloons into the air, and earlier this century, most passenger-carrying balloons were filled with hydrogen. In the 1920s and 1930s, luxurious airships were built to carry passengers across the Atlantic with all the grace and style of ocean liners. These huge aircraft were lifted by rigid, cigar-shaped bags full of hydrogen gas. Once airborne, propeller engines drove the airship along. Unfortunately, hydrogen is highly flammable. In 1937, the *Hindenburg*, one of the biggest and grandest of airships—over 800 ft (245 m) long—was engulfed in flames in a terrible accident. As the airship was coming in to land at an airfield in New Jersey, it collided with a metal pylon. Within seconds, the balloon was a fireball. Amazingly, there were survivors. However, the tragedy spelled the end for airships since 35 passengers—and one person on the ground—died that day.

Burning

Hydrogen is one of the most highly flammable of all gases. Hydrogen gas burns in air to produce water. Two diatomic molecules of hydrogen combine with one diatomic molecule of oxygen to form two molecules of water. The burning may be a gentle, squeaky "pop," or perhaps a huge explosion with roaring flames. How dramatic the burning is depends on how the hydrogen mixes with the air.

If hydrogen is allowed to emerge in a steady jet from a gas cylinder, it usually burns quietly. The barely visible flame is pale blue, perhaps tinged with yellow by the forming drops of water. But when hydrogen and oxygen mix in the right proportions—about two to one—the mixture can violently explode. So hydrogen is a potentially dangerous gas that must be handled with care.

Reducing power

Any substance that draws oxygen out of a compound is called a reducing agent, and hydrogen is one of the strongest known. In fact, it will take the oxygen out of nearly all metal oxides if they are heated enough. For example, heated copper oxide reacts with hydrogen to form copper and steam. How strongly hydrogen reduces an oxide depends on where the metal is in the Reactivity Series—a table showing how readily each metal reacts.

In 1937, the airship Hindenburg *crashed as it was coming in to land in New Jersey, and the flammable hydrogen gas inside exploded into flames. Remarkably, only 35 of the 97 people on board the airship were killed in this tragic accident.*

Hydrogen will also reduce natural oils, taking the oxygen out of them and transforming them into solid fats. This reaction is important in the manufacture of margarine. Hydrogen is used to take the oxygen out of vegetable oils and turn them into solid margarine.

Hydrogen ions

Because they have just one electron, hydrogen atoms have a "gap" in their outer shell, making them unstable. What makes hydrogen special is that its atoms can form both positive and negative ions. They can either lose an electron to form a positive ion, or they can gain one to form a negative ion.

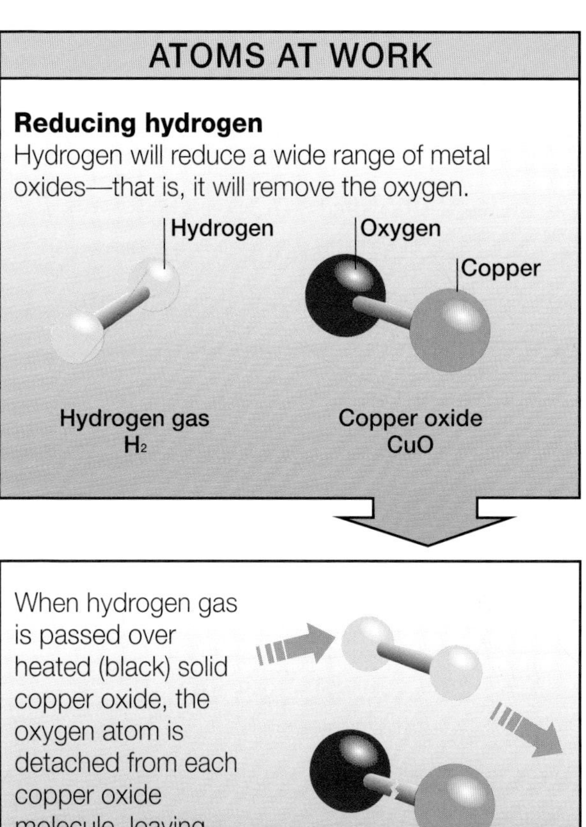

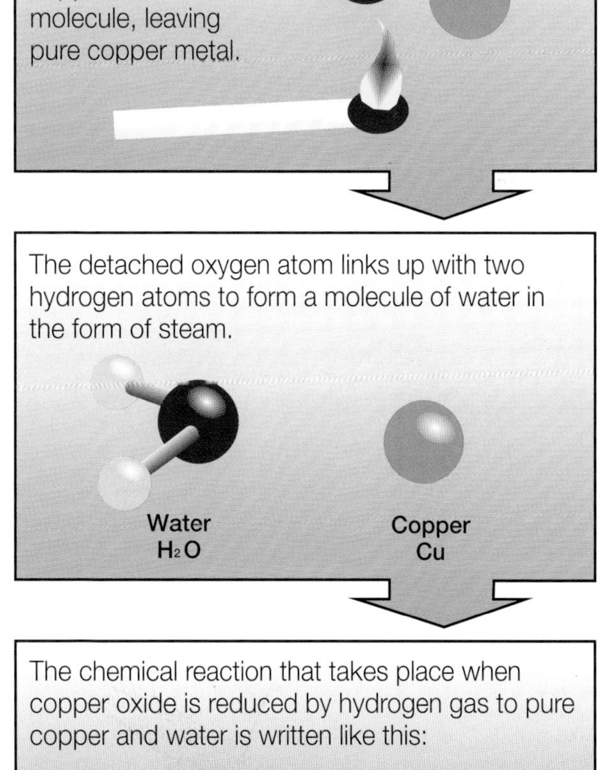

Positive hydrogen ions can be formed in the laboratory using discharge tubes, which are a bit like television tubes. More often, they are found in watery liquids (called aqueous solutions). Here they do not form in isolation, but link up with water molecules, forming what are known as hydronium ions, which have the formula H_3O^+. Positive hydrogen ions will react with a whole range of compounds. They react with oxides and hydroxides, for example, to form water; with carbonates to form water and carbon dioxide; and with sulfides to form hydrogen sulfide.

Negative hydrogen ions—a proton and two electrons—form when hydrogen comes in contact and reacts with a wide range of metals, such as calcium.

Hydrogen and metals

Hydrogen gas will react with many hot metals to form white "hydride" crystals that look like table salt. These hydrides, known as ionic hydrides, are formed with the metals in the two groups on the left of the periodic table—lithium, sodium, potassium, rubidium, caesium, calcium, strontium, and barium. When these ionic hydrides come in contact with water, they react vigorously to give hydrogen. Some of the more complicated hydrides, such as lithium tetrahydroaluminate (the hydride of lithium with aluminum) are very valuable as reducing agents.

Metals with more electrons in the outer rings of their atoms, such as iron and copper, can absorb huge quantities of hydrogen without basically changing. The hydrogen seeps into the metal's structure rather like water into a sponge. The tiny hydrogen atoms are dotted throughout the structure but do not have much of an effect on the metal's character. Hydrides that form in this way are known as "interstitial" or metallic hydrides.

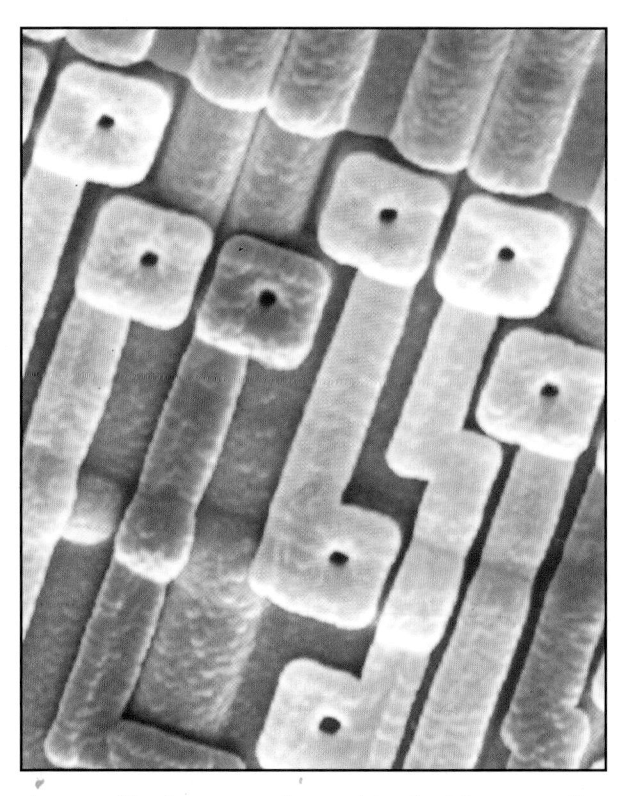

Atoms of hydrogen can be combined with atoms of the nonmetal silicon to form a very useful compound called silane. Silane is a gas and is invisible. Silicon in a more visible form, as a solid, is an essential constituent of computer microchips such as the one shown here.

This man is welding using a torch that burns acetylene together with oxygen. Acetylene is one of a group of hydrogen compounds called hydrides. Most hydrides, including acetylene, are gases.

Hydrogen and nonmetals

Not all of hydrogen's reactions involve ions. A hydrogen atom can also fill its outer shell by joining with another atom that has a gap or gaps in its outer shell. Atoms that share electrons this way form what is termed a covalent bond.

DID YOU KNOW?

SILANE

An extremely useful covalent hydride is the gas silane, the hydride of silicon. Silicon used in silicon chips for the computer industry is obtained from this colorless gas. Silane is also used to clean out corroded pipes in nuclear reactors. Silane is made by combining the ionic hydride, lithium tetrahydroaluminate, with silicon tetrachloride, or from the reaction of magnesium silicide with acid.

Hydrogen can form covalent bonds with virtually all nonmetals except for the noble gases. Other than water and hydrogen fluoride, the resulting covalent hydrides that are formed are all gases at room temperature.

When hydrogen combines directly with a nonmetal, the proportion that reacts is called the "yield." The yield is often small. The yield of hydride can be improved if the nonmetal is combined with hydrogen atoms in the form of a dilute acid, and the mixture is then heated. For example, when calcium carbide is mixed with water or methane (a compound of carbon and hydrogen) and this mixture is heated to 2,732°F (1,500°C), the hydride acetylene, used in welding, is formed.

Isotopes

The basic hydrogen atom—a single proton, circled by a single electron—is also called protium. In 1919, British physicist Francis Aston (1877–1945) made a strange discovery while studying hydrogen. At the time, measurements put the atomic mass of hydrogen at 1.00777, yet Aston's measurements gave a figure of 1.00756. Most researchers assumed that the earlier samples of hydrogen, with the mass of 1.00777, must had been contaminated by a heavier, unknown substance.

In 1932, U.S. physicist Harold Urey (1893–1981) found that the heavier atoms were not another substance but a different form, or isotope, of the hydrogen atom. This heavy atom is called deuterium. Like protium, it has one proton and one electron, so it still has all of the chemical properties of hydrogen. But the deuterium nucleus also contains a neutron, giving the atom its extra mass.

Heavy water

Urey found that deuterium occurs in tiny amounts in ordinary water. When cosmic rays from space hit the oceans, a tiny proportion of water molecules have their protium atoms changed into deuterium.

Urey purified this form of water, called heavy water, by electrolysis. In this process, an electric current is passed through water, splitting its molecules into hydrogen and

British scientist Francis Aston won a Nobel Prize for his work on the isotopes of hydrogen.

DID YOU KNOW?

GETTING HEAVY

At first, making heavy water was so tricky and used so much energy that it was extremely rare and valuable. In 1940, at the beginning of World War II, the only significant stock of heavy water—53 gallons (200 liters), about a bathful—was in Norway. Yet Norway was in danger of Nazi invasion. Rather than let this precious stock fall into Nazi hands, French physicist Frédéric Joliot-Curie (1900–1958) had the water smuggled to Britain by submarine. The Norwegian heavy-water plant was then blown up. With nearly all the world's stock of heavy water, the Allies had a huge advantage in the race to develop the atom bomb.

ATOMS AT WORK

Ordinary hydrogen atoms, called protium and written H, have one proton in their nucleus, circled by a single electron.

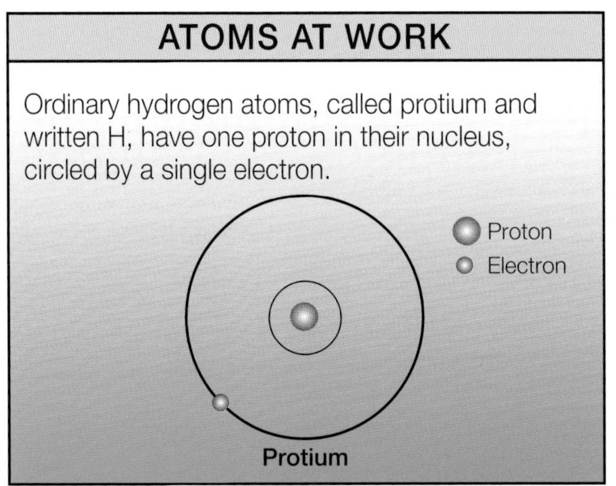

Deuterium atoms (written ^{2}H, or D) have one proton and one neutron in their nucleus, circled by a single electron.

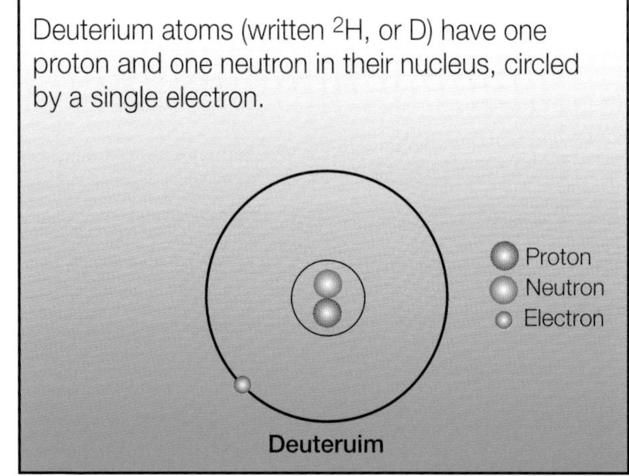

Tritium atoms have one proton and two neutrons in their nucleus, circled by a single electron.

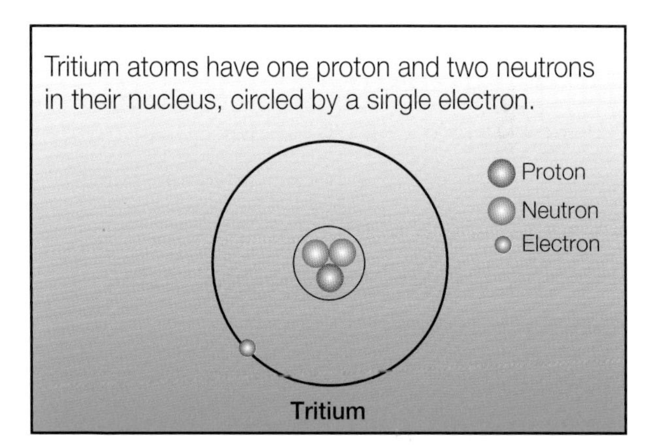

oxygen. Molecules of heavy water break up more slowly than ordinary water, so the longer electrolysis goes on, the richer in deuterium the water becomes. Finally, no protium atoms are left—only pure heavy water, called deuterium oxide or D_2O.

Heavy water is very good at absorbing neutrons. This makes it very useful in controlling nuclear reactions. It played a crucial role in the development of the first atomic bomb. Today, many nuclear power stations use heavy water to prevent their reactors from melting down.

The heaviest hydrogen of all

In 1934, scientists found a third form of hydrogen—tritium. As well as a proton and an electron, each atom of tritium has two neutrons. It is three times heavier than protium and is very unstable. Most tritium is made in nuclear reactors, but tiny amounts are made when cosmic rays bombard water vapor in the atmosphere.

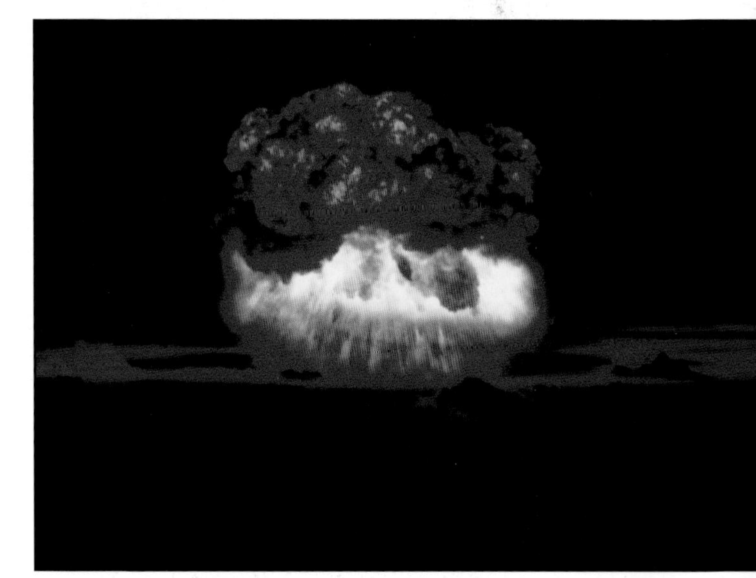

This controlled nuclear explosion occurred on February 28, 1954 on the desert surface in Nevada.

Uses of hydrogen

Hydrogen is an incredibly useful element, and it can be used in a huge variety of ways. In essence, though, hydrogen is used in four main ways—as a reducing agent to modify and purify chemicals, as a fuel, in acids, and as a constituent of various other compounds.

Reducing hydrogen

Many metals do not occur naturally in the ground in their pure forms. Instead, they are extracted as metallic oxide ores. To refine these oxide ores into pure metal, they have to be reduced. Because it is a good reducing agent, hydrogen can withdraw the oxygen and other nonmetallic compounds from the oxide ore, leaving a residue of pure metal.

At low temperatures, hydrogen is a much better reducing agent than carbon, its only real rival. Above 1,300°F (700°C), however, carbon works better and is easier to use. So in industry, most metals are purified using carbon at high temperatures. However, there are some metals, including tungsten, molybdenum, and nickel, that are more easily reduced using hydrogen. If a stream of hydrogen is passed over heated tungsten trioxide, for example, the hydrogen removes the oxygen and combines with it to form water, leaving pure tungsten metal.

These lightbulbs all contain tungsten filaments, which improve the bulbs' efficiency. Hydrogen is important in the manufacture of tungsten. It is a powerful reducing agent and it produces tungsten by reducing the metal from its oxide.

Hydrogenation

One particularly important use of hydrogen as a reducing agent is in turning liquid oils to solids. The margarine you spread on bread is not made from milk like butter is, but from vegetable oils such as sunflower oil. These liquid oils are made from molecules that are essentially combinations of carbon and hydrogen atoms. To turn them into solid fats, they are mixed with hydrogen. Extra hydrogen atoms join on to the oil molecules, turning the oil to solid fat. This process is called hydrogenation. By controlling the amount of hydrogen added, margarine can be hard or soft, as required.

Margarine and other low-fat spreads are not the only substances made this way. Cyclohexane, the solvent used in many types of paint remover, is made by hydrogenating benzene (a compound that consists of six carbon atoms and six hydrogen atoms). Oil can be made from coal dust by hydrogenating coal powder with a little oil under heat and pressure. Hydrogenation is also important in the making of gasoline from crude petroleum.

On the other hand, acetone, which is used as a solvent in many cosmetics (such as nail-varnish remover), is made by the opposite process—called dehydrogenation—that is, the oil propanol is thinned by losing two of the eight hydrogen atoms in its molecule.

DID YOU KNOW?

WORTH A NICKEL

You would not have margarine without nickel. The hydrogenation reaction is only really efficient if there is another chemical, called a catalyst, to help it on its way. In 1912, French chemist Paul Sabatier (1854–1941) discovered that nickel is a good hydrogenation catalyst and received the Nobel Prize for his efforts. Other hydrogenation catalysts used today include platinum and Raney nickel—a special alloy (mixture) of nickel and aluminum. Raney nickel is used as the catalyst for making paint remover from benzene.

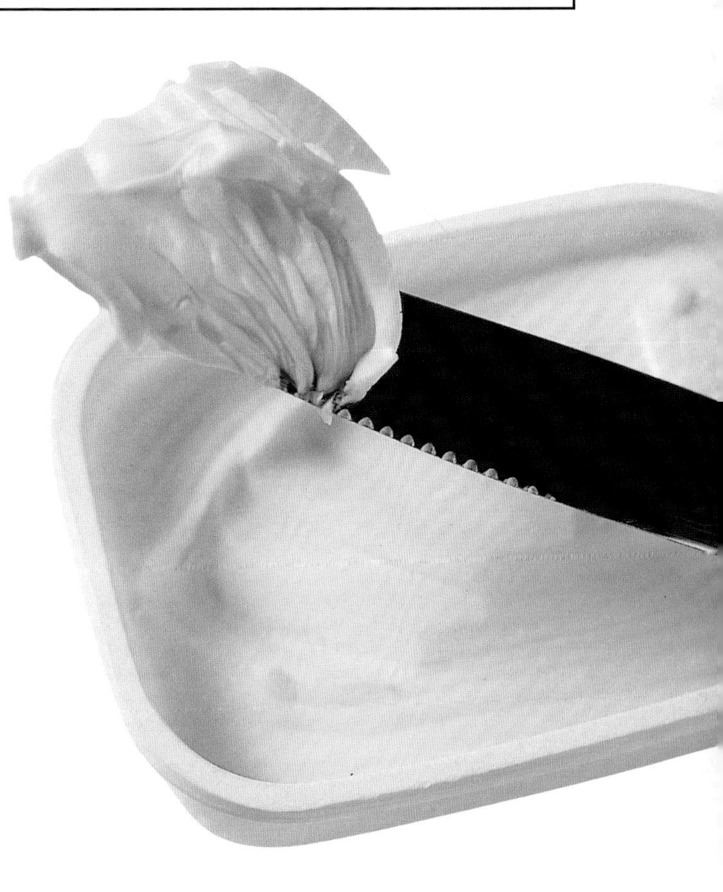

Hydrogen is important in the manufacture of everyday things such as the margarine shown here.

Hydrogen compounds

Hydrogen forms so many useful compounds that it is only possible to mention a few of the better-known ones. When combined with nitrogen, for example, it forms ammonia, which is an essential ingredient of many artificial fertilizers. The ammonia is usually used to make compounds called nitrates, which are spread on the soil as powder. Sometimes, however, liquid ammonia is poured directly on to the soil. The nitrogen in the ammonia helps plants to grow.

Another useful hydrogen compound is hydrogen peroxide. Each molecule of hydrogen peroxide consists of two hydrogen atoms and two oxygen atoms, giving it a chemical formula of H_2O_2. It is a colorless, syrupy liquid with a sharp scent. At high concentrations, it can be very corrosive (meaning that it eats into materials) and can burn skin badly. But diluted with water, it can be used as an antiseptic to kill germs, and as a skin cleaner. Hydrogen peroxide is also a very good bleach. It is used for bleaching hair blonde and to make paper and cotton white. Without it, your paper, shorts, and shirts would not be sparkling white but a dingy brown!

A better known hydrogen compound, though, is probably ethyl alcohol. This is a compound of hydrogen, oxygen, and carbon. It is the alcohol in alcoholic drinks such as beer and wine.

ATOMS AT WORK

Making ammonia

Ammonia is made using hydrogen and nitrogen in the Haber process. Normally, these two gases exist as molecules made of pairs of atoms held together by strong bonds.

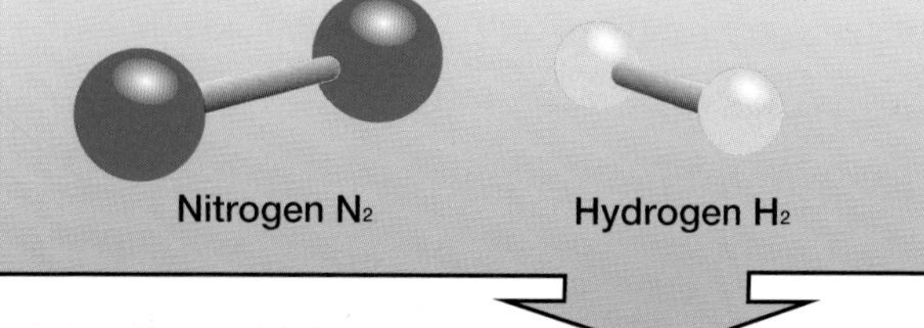

When they are mixed and compressed, the bonds in the nitrogen and hydrogen molecules break apart into separate atoms.

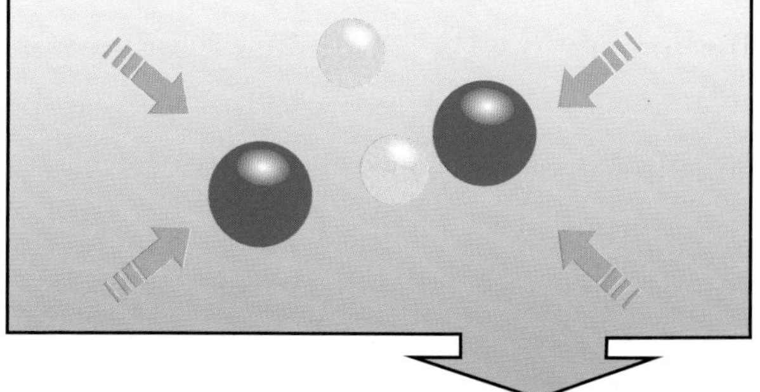

When the compressed mixture is passed over iron in a heated catalyst chamber, one in ten nitrogen atoms combines with three hydrogen atoms to form an ammonia molecule.

Ammonia NH_3

The chemical reaction that takes place in the Haber process is written like this:

$$N_2 + 3H_2 \rightarrow 2NH_3$$

This tells us that one molecule of nitrogen combines with three molecules of hydrogen to give two molecules of ammonia.

Making hydrogen

Hydrogen can be made in dozens of different ways. In the laboratory, it can be made by the reaction of certain metals with acids. If zinc metal is placed in a beaker of dilute hydrochloric acid, bubbles of hydrogen are formed. Some very reactive metals, such as sodium, react with water to produce hydrogen.

In the past, most hydrogen was made industrially by a reaction called the Bosch process. In this process, steam is passed over red-hot coke to make a mixture of carbon monoxide and hydrogen, called water gas. The water gas is then mixed with more steam at very high temperatures in the presence of, first, iron and chromium and, second, copper catalysts. The carbon monoxide in the water gas is converted to carbon dioxide, which can then be removed by dissolving in water, leaving behind the hydrogen.

> **DID YOU KNOW?**
>
> ***NEW SOURCES OF HYDROGEN***
>
> At present, about 30 million tons (27 million tonnes) of hydrogen are made worldwide every year. Scientists are looking for new ways to make this useful element. One possibility is the idea of using heat-loving bacteria. In 1996, American scientists at the Oak Ridge National Laboratory in Tennessee discovered that some bacteria, found living on coal slag heaps and inside hydrothermal volcanic vents deep beneath the sea, could make hydrogen from glucose. Using these bacteria, it may be possible to make huge quantities of hydrogen at very little cost from wood and waste paper.

Today most hydrogen is made from natural gas. Steam is passed through natural gas (methane) to create a mixture of carbon monoxide and hydrogen, called synthesis gas or syngas, which is purified in the same way as water gas.

This automobile is being filled with gasoline. Gasoline is part of the hydrocarbon group of compounds, which consist only of hydrogen and carbon atoms.

The power of hydrogen

Our supplies of fossil fuels, such as oil and natural gas, are being used up at alarming rates—in power stations, automobiles, and airplanes, while the pollution caused by their burning is damaging the environment.

DID YOU KNOW?

ROCKET POWER

Hydrogen has been a principal fuel for rockets since the 1930s. It was used in the German V-2 rockets during World War II. Now, many space rockets carry tanks of liquid hydrogen and liquid oxygen. The two liquids are released together and burned. As they expand, they create a very powerful thrust that drives the rocket upward. Hydrogen and oxygen liquid fuels have been used to power everything from the upper stages of the *Saturn V Apollo* rocket to the main engines of the space shuttle. There is a store of 800,000 gallons (3 million liters) of liquid hydrogen in just one of the storage tanks at the U.S. space program headquarters at Cape Canaveral, Florida.

Scientists are looking for alternatives to fossil fuels—and hydrogen is one of the most promising. A big advantage of using hydrogen is that when it burns, it does not produce soot, carbon dioxide, carbon monoxide, or any of the other noxious substances left by fossil fuels. It simply combines with oxygen in the air to create water vapor. The world it leaves may be damp, but it will also be clean.

Flames pour from the engines of the space shuttle Discovery, *which is fueled by liquid hydrogen.*

There are now plans to supply homes with hydrogen once natural gas runs out, and hydrogen-powered atomobiles and airplanes are now well beyond the drawing board. In 1992, an automobile that runs on hydrogen was built in Japan by Nissan. By 1996, BMW, the German automobile company, had six hydrogen-powered automobiles on the road.

Storage of liquid hydrogen can be a problem. Safer than having tanks of liquid hydrogen is to carry the hydrogen around absorbed in certain metal alloys—like water in a sponge. Mazda has built an automobile that stores hydrogen in this way, but instead of burning the hydrogen as a fuel, it is used to provide electrons, generating electricity to run the car. Now, Daimler-Benz Aerospace Airbus is working on a hydrogen-powered aircraft.

DID YOU KNOW?

HYDROGEN TANKS

One problem with hydrogen as a fuel is how to store it. As a gas, 1 lb. (0.454 kg) of hydrogen takes up over 1,450 gallons (5,500 liters). But if it is turned into a liquid, it occupies just 1.5 gallons (5.5 liters). And it gives three times as much energy as the same volume of gasoline! To stay liquid, hydrogen must be kept cold and under high pressure. In the BMW hydrogen-powered automobile, liquid hydrogen is kept in strong metal containers under a pressure five times that of normal atmospheric pressure. If carrying liquid hydrogen around seems dangerous, it is no more so than gasoline.

Fitting snugly inside the trunk of this BMW test automobile is a fuel tank filled not with gasoline but with liquid hydrogen. In the picture, the tank is being filled at a special service station.

Hydrogen and acids

There are many different acids, from the tartaric acid in grapes to the sulfuric acid in car batteries. A weak acid called citric acid makes lemons taste sour. Stronger acids like hydrochloric acid can dissolve away the strongest steel.

What all acids have in common is that they contain hydrogen. Substances only become acids when they dissolve in water, allowing the hydrogen to break away as positively charged ions—that is, hydrogen atoms that have lost their electrons.

The number of hydrogen ions an acid can form in water is a measure of its strength, or pH. The pH scale goes from 0 to 14, and the lower the pH, the more acidic a substance is. A neutral substance has a pH of 7. Anything with a pH of more than 7 is alkaline. Anything with a pH of less than 7 is acidic.

In some strong acids, such as nitric and sulfuric acid, the original molecules split apart completely in water so that there are huge numbers of hydrogen ions floating

Lemon juice turns blue litmus paper red, showing that lemons contain acid (citric acid).

SEE FOR YOURSELF

MAKING AN ACID TESTER

Chemists use indicators, such as litmus paper and phenolphthalein, to test acidity. Indicators change color according to whether a substance is acid or alkali. You can make your own indicator from red cabbage, elderberries, or blackberries.

- To make the indicator, chop a red cabbage and boil it in distilled water—you must use distilled water—then leave it to cool.
- Collect some things to test. You could try such things as: lemon or lime juice, bottled water, distilled water, salt water, milk, vinegar, baking soda, indigestion tablets, disinfectant, or bleach.
- Put a little red-cabbage water in a test jar (perhaps an old herb jar). Add a little of the substance to be tested.
- If the water turns red, the substance is an acid. If it stays purple, the substance is neutral. If it goes blue, the substance is a weak alkali. If it goes green, the substance is a strong alkali.

free. In weaker acids, such as citric acid, only a few molecules break up, releasing fewer hydrogen ions. Acids can be diluted by adding more water, so that the hydrogen ions are spread more thinly, or they can be concentrated by evaporating some of the water. A dilute strong acid and a concentrated weak acid may have close to the same pH.

Acids attack metals because the hydrogen ions, with their missing electrons, grab electrons from the metal atoms. As a result, a salt of the metal forms and hydrogen gas is released. An acid reacts with a base to form a salt and water.

How acids are used

Acids are found in many places and have many uses. Strong acids make good electrolytes—that is, liquids that conduct electricity. This is because they are almost entirely split up into positive hydrogen ions

DID YOU KNOW?

The indicator litmus has been known about for over 400 years. It is a blue substance that comes from lichens that grow on clifftops and on trees in warm climates. It turns red in the presence of an acid but turns blue again when an alkali is added.

This model represents the structure of a molecule of deoxyribonucleic acid (DNA). This acid is the basic building block of life.

One acid that most people would prefer we did not have is acid rain. Acid rain causes damage to buildings, plants, and animals. In this picture, lime—an alkaline substance—is being added to a stream in Sweden to neutralize the effects of the acidity in the water.

and negative non-hydrogen ions. This abundance of ions, all of which are electrically charged, carries an electric current well. This is why sulfuric acid is used in many car batteries. Phosphoric acid makes a good rust remover. It reverses the reaction between iron and oxygen that causes rust to form.

ACID POWER

pH	
0.0	Concentrated hydrochloric acid
1.5	Stomach acid
2.8	Lemon juice
3.2	Vinegar
4.0	Fruit juices
4.5	Soda water
5.5	Rainwater
6.7	Fresh milk
7.0	Pure water
7.8	Blood
8.8	Seawater
11.0	Ammonia solution
12.2	Lime water
14.0	Sodium hydroxide solution

Weak acids, on the other hand, play an important role in many living processes. Indeed, the molecule that carries the information needed to make a living organism, the DNA inside most living cells, is an acid. DNA is short for deoxyribonucleic acid.

Another group of acids vital for life is the amino acids, the building blocks from which all proteins are made. A fairly strong acid, hydrochloric acid, in our stomachs helps us to digest our food.

Acids can also be dangerous to some forms of life, and this can be useful to us. In the pickling process, vinegar (ethanoic or acetic acid) is used to preserve food by preventing bacteria and molds from being able to grow on it.

Hydrogen bonds

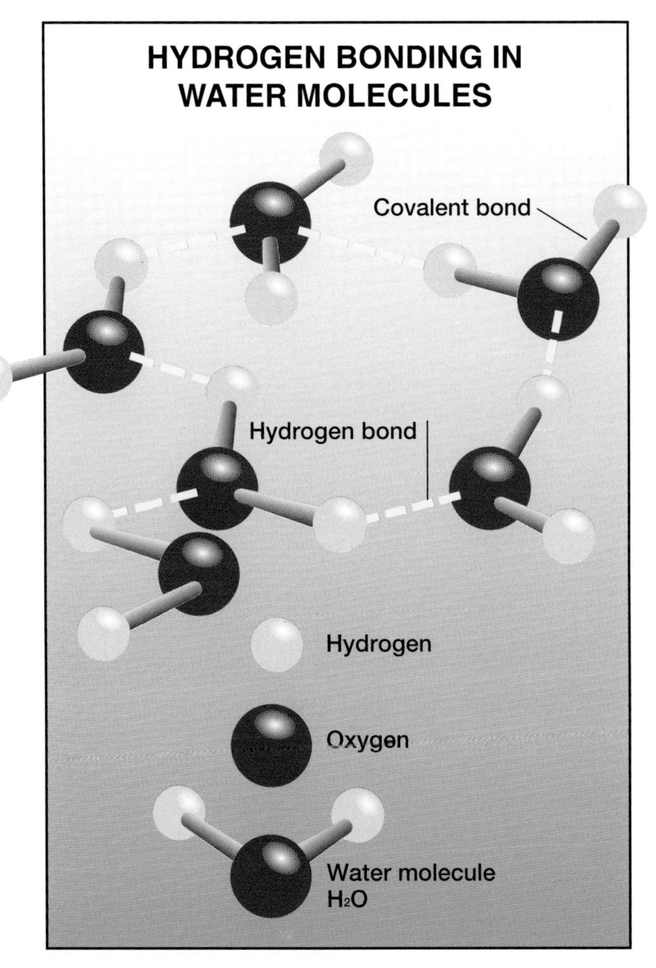

Because the hydrogen nucleus is so small and light, it does not attract electrons very strongly. In other words, it is very weakly electronegative, and this has important effects. Most other nonmetallic elements are more electronegative than hydrogen—that is, they attract electrons to them more strongly than hydrogen does. The most powerfully electronegative elements are oxygen and the halogen gas fluorine, but even nitrogen is more strongly electronegative than hydrogen.

When a hydrogen atom joins covalently to a more strongly electronegative atom, to form a molecule, the electrons in the molecule are always drawn toward the other atom more strongly than they are toward the hydrogen atom. In water molecules (two hydrogen atoms and one oxygen atom), electrons are drawn toward the oxygen atom and cluster around it. In ammonia molecules (three hydrogen atoms and one nitrogen atom), electrons are attracted to the nitrogen atom.

Since electrons carry a negative charge, the effect is to polarize the molecule. This means that one end of the molecule becomes negatively charged and the other end becomes positively charged. These charged molecules are called polar molecules. They behave like tiny magnets.

Since opposite charges attract, the negative end of one polar molecule is attracted to the positive end of another polar molecule. The effect of this mutual attraction is to form a weak bond between the two molecules.

When polar molecules involving hydrogen link up like this, the attraction between them is called a hydrogen bond. The hydrogen bond is an example of a particularly strong attraction between polar molecules, and it produces some unusual effects.

Water

Because water, when it freezes as the solid ice, is less dense than it is as a liquid—and less dense substances always float on denser ones—these icebergs float on the sea.

Hydrogen bonding helps to explain some of the unique properties that make water so suitable for life.

One remarkable thing about water is that it is a liquid (rather than a gas) at normal temperatures. Considering that both oxygen and hydrogen atoms are very lightweight—oxygen and hydrogen are both gases at temperatures very far below freezing—you might expect water to be a gas, too. Water is a liquid because the hydrogen bonds formed between the water molecules hold the molecules together until the temperature approaches 212°F (100°C)—called the boiling point. In fact, the hydrogen bonds in water are 10 times stronger than the bonds between molecules in a substance like wax.

DID YOU KNOW?

COLD FACTS

Because ice is less dense than water, ice floats, which explains why we are able to see icebergs. Ice not only floats, it actually forms on top of the water. As water cools below 39°F (4°C), it begins to expand. So, instead of sinking, cold water actually rises to the surface. If cold water sank, the oceans would gradually freeze from the bottom up—and would probably never melt. On a smaller scale, this is crucial for pond life. No ice forms on a pond until all the water has cooled to 39°F (4°C). When the pond does freeze, it only freezes on top. Beneath the ice, life can go on at a steady temperature of 39°F (4°C).

When water is heated to its boiling point, the hydrogen bonds that hold the liquid together gradually loosen, until the water molecules escape as water vapor. On contact with the cooler atmosphere, water vapor condenses back into liquid, which we can see as steam. The hydrogen bonds do not disappear altogether—water vapor molecules cluster together much more than other gas molecules. This is why you can usually see the steam coming from a kettle and why water vapor in the atmosphere so readily forms into clouds.

Hydrogen bonds are responsible for another unusual property of water, the fact that it expands as it freezes. The crystalline structure of ice holds the water molecules firmly in place, preventing hydrogen bonding from working at its maximum. As ice melts, the structure breaks down, and the water molecules, attracted by the power of hydrogen bonding, pack closer together. So water fills less space and is denser than ice. In fact, water is at its most dense at 39°F (4°C), several degrees above its melting point. Conversely, when water freezes, it expands and becomes lighter.

While we take it for granted, life without this property would be very difficult indeed. Still, there are some drawbacks—such as burst pipes in winter. But if water did not expand like this, Earth's oceans would slowly freeze over until our planet became a giant ball of ice.

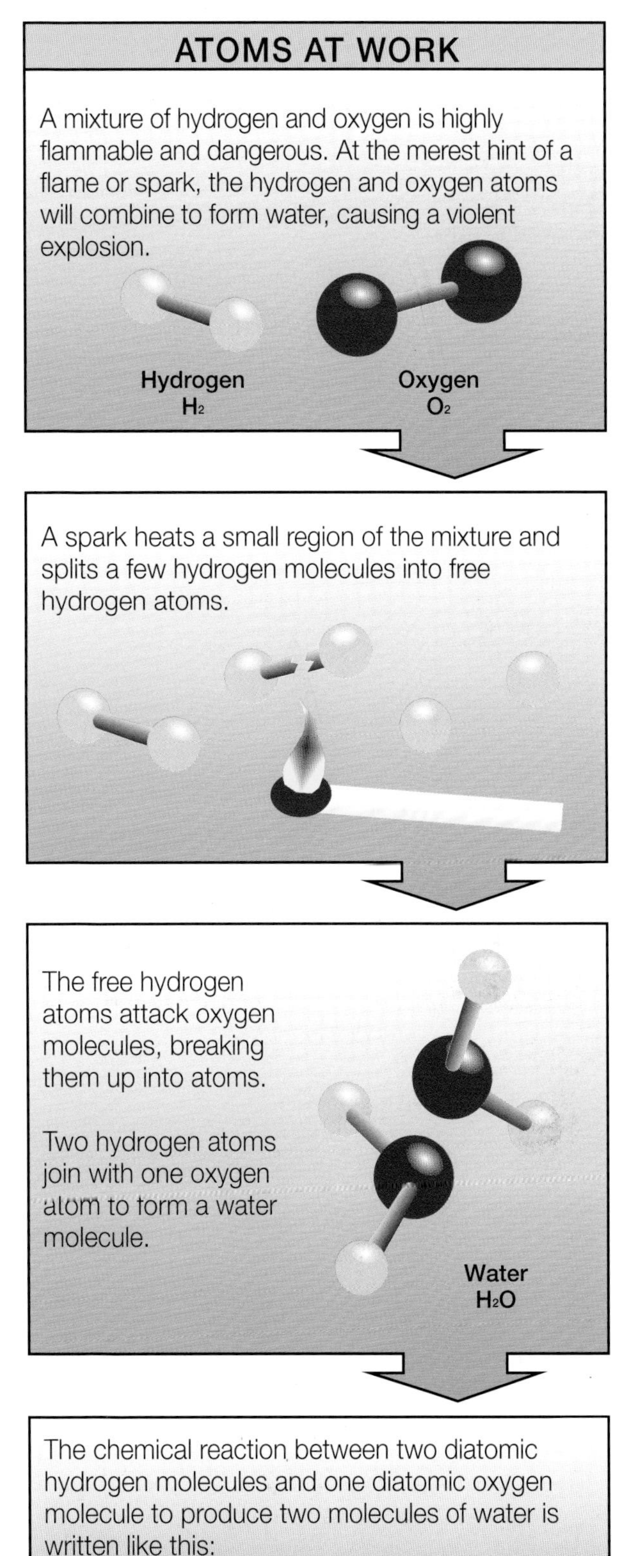

Periodic table

Everything in the universe is made from combinations of substances called elements. Elements are the building blocks of matter. They are made of tiny atoms, which are much too small to see.

The character of an atom depends on how many even tinier particles called protons there are in its center, or nucleus. An element's atomic number is the same as the number of protons.

Scientists have found around 110 different elements. About 90 elements occur naturally on Earth. The rest have been made in experiments.

All these elements are set out on a chart called the periodic table. This lists all the elements in order according to their atomic number.

The elements at the left of the table are metals. Those at the right are nonmetals. Between the metals and the nonmetals are the metalloids, which sometimes act like metals and sometimes like nonmetals.

- On the left of the table are the alkali metals. These elements have just one electron in their outer shells.
- On the right of the periodic table are the noble gases. These elements have full outer shells.
- Elements in the same group have the same number of electrons in their outer shells.
- Elements get more reactive as you go down a group.
- The number of electrons orbiting the nucleus increases down each group.
- The transition metals are in the middle of the table, between Groups II and III.

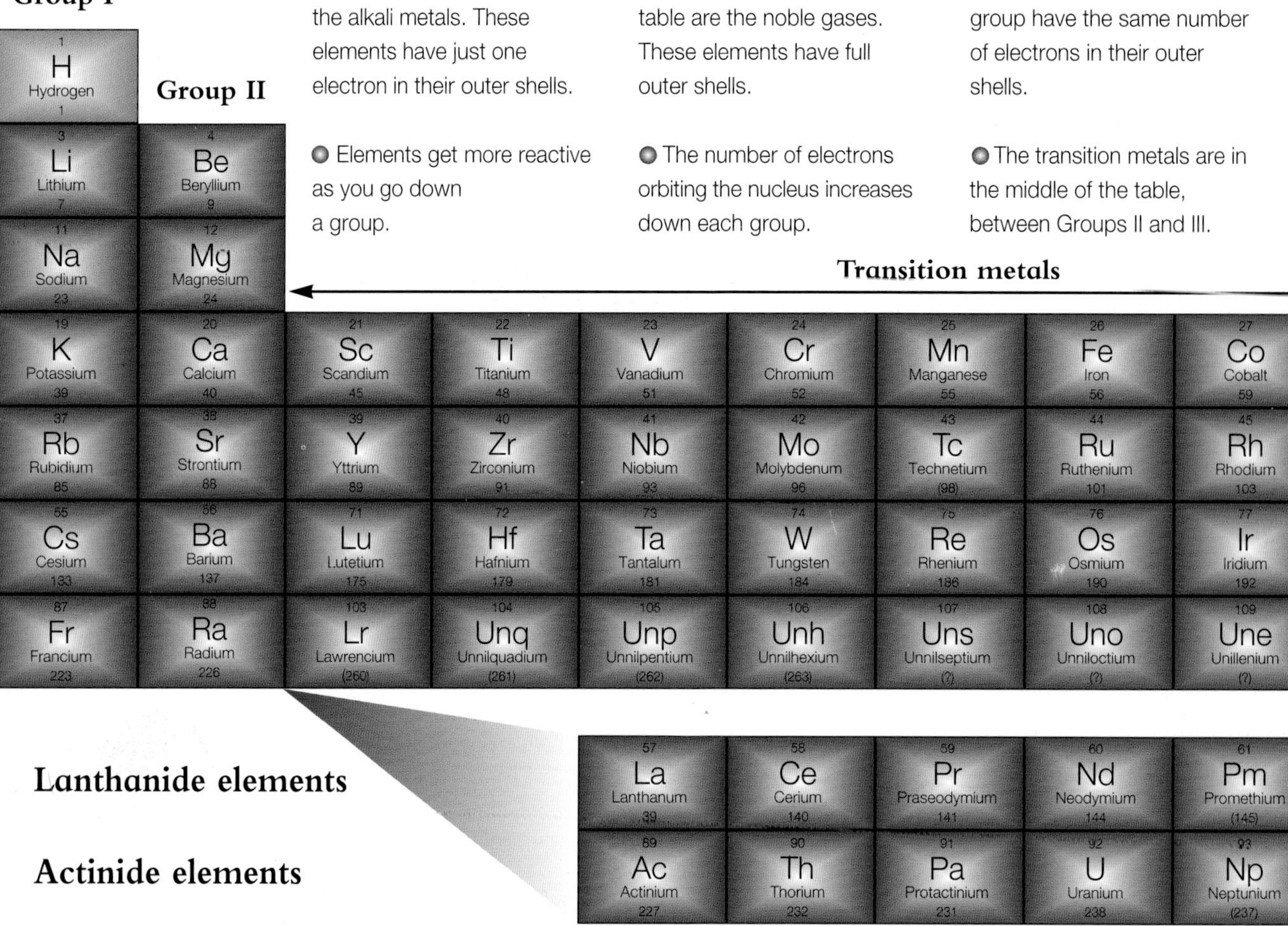

The horizontal rows are called periods. As you go across a period, the atomic number increases by one from each element to the next. The vertical columns are called groups. Elements get heavier as you go down a group. All the elements in a group have the same number of electrons in their outer shells. This means they react in similar ways.

The transition metals fall between Groups II and III. Their electron shells fill up in an unusual way. The lanthanide elements and the actinide elements are set apart from the main table to make it easier to read. All the lanthanide elements and the actinide elements are quite rare.

Hydrogen in the table

Hydrogen is the lightest of all the elements, with an atomic number of one, so it has only one proton in its nucleus. It is in Group I, together with the highly reactive "alkali" metals. At normal temperatures, hydrogen is a highly volatile gas. It has the lowest boiling point of any element in the periodic table.

Metals
Metalloids (semimetals)
Nonmetals

Key	
1	Atomic (proton) number
H	Symbol
Hydrogen	Name
1	Atomic mass

			Group III	Group IV	Group V	Group VI	Group VII	Group VIII
								2 He Helium 4
			5 B Boron 11	6 C Carbon 12	7 N Nitrogen 14	8 O Oxygen 16	9 F Fluorine 19	10 Ne Neon 20
			13 Al Aluminum 27	14 Si Silicon 28	15 P Phosphorus 31	16 S Sulfur 32	17 Cl Chlorine 35	18 Ar Argon 40
28 Ni Nickel 59	29 Cu Copper 64	30 Zn Zinc 65	31 Ga Gallium 70	32 Ge Germanium 73	33 As Arsenic 75	34 Se Selenium 79	35 Br Bromine 80	36 Kr Krypton 84
46 Pd Palladium 106	47 Ag Silver 108	48 Cd Cadmium 112	49 In Indium 115	50 Sn Tin 119	51 Sb Antimony 122	52 Te Tellurium 128	53 I Iodine 127	54 Xe Xenon 131
78 Pt Platinum 195	79 Au Gold 197	80 Hg Mercury 201	81 Tl Thallium 204	82 Pb Lead 207	83 Bi Bismuth 209	84 Po Polonium (209)	85 At Astatine (210)	86 Rn Radon (222)

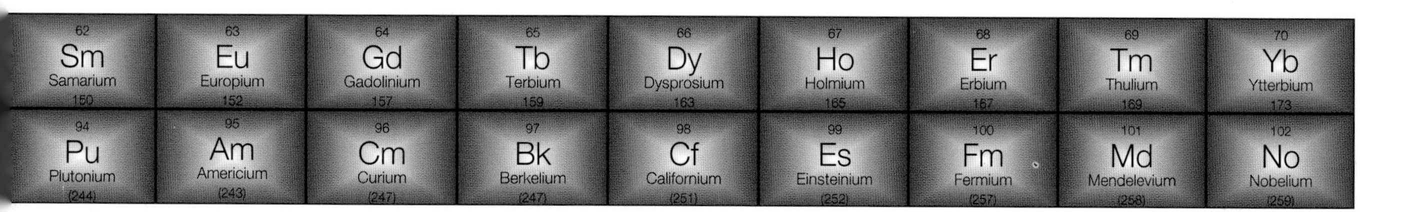

62 Sm Samarium 150	63 Eu Europium 152	64 Gd Gadolinium 157	65 Tb Terbium 159	66 Dy Dysprosium 163	67 Ho Holmium 165	68 Er Erbium 167	69 Tm Thulium 169	70 Yb Ytterbium 173
94 Pu Plutonium (244)	95 Am Americium (243)	96 Cm Curium (247)	97 Bk Berkelium (247)	98 Cf Californium (251)	99 Es Einsteinium (252)	100 Fm Fermium (257)	101 Md Mendelevium (258)	102 No Nobelium (259)

Chemical reactions

Chemical reactions are going on around us all the time. Some reactions involve just two substances; others many more. But whenever a reaction takes place, at least one substance is changed.

In a chemical reaction, the atoms stay the same. But they join up in different combinations to form new molecules.

Ammonium nitrate burning. This compound is manufactured on a large scale from ammonia and is used to make fertilizers and explosives.

ATOMS AT WORK

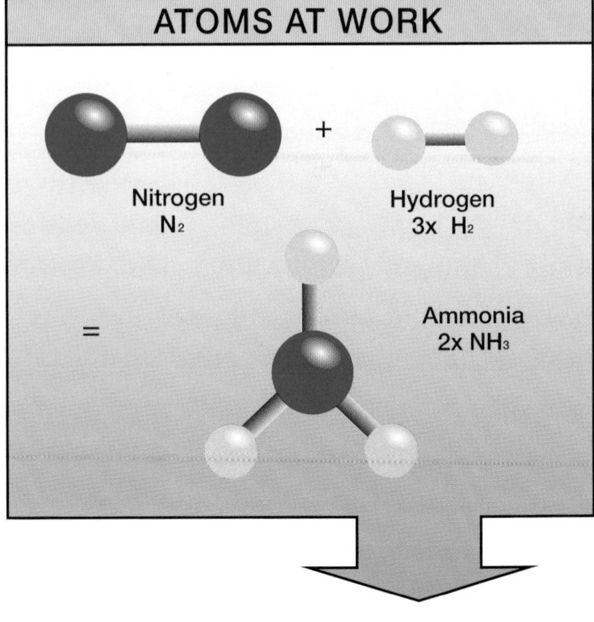

The chemical reaction that takes place when nitrogen and hydrogen are mixed together to form ammonia is written like this:

$$N_2 + 3H_2 \rightarrow 2NH_3$$

Writing an equation

Chemical reactions can be described by writing down the atoms and molecules before and the atoms and molecules after. As the atoms stay the same, their number before is the same as their number after. Chemists write reactions as equations.

Making it balance

When the numbers of each atom on both sides of the equation are equal, the equation is balanced. If the numbers are not the same, something is wrong. The chemist adjusts the number of atoms until the equation balances.

Glossary

acid: A substance that can provide hydrogen atoms for chemical reactions.
acid rain: When certain gases rise into the atmosphere, they dissolve in rainwater, making the rain acidic.
alcohol: A compound that contains hydrogen, carbon, and oxygen.
alkali: A compound that dissolves in water to give a solution with a pH greater than 7.
atom: The smallest part of an element that still has all the properties of that element.
atomic number: The number of protons in an atom.
bond: The attraction between two atoms that holds the atoms together.
compound: A substance that is made of atoms of more than one element. The atoms in a molecule are held together by chemical bonds.
deuterium: An isotope of hydrogen with one neutron in the nucleus.
electron: A tiny particle with a negative charge. Electrons are found inside atoms, where they move around the nucleus in layers called electron shells.
fossil fuels: Fuels including coal, oil, and natural gas that formed from the bodily remains of prehistoric plants and animals.
heavy water: Water that contains deuterium attached to oxygen.
hydride: A compound of hydrogen with another element.
isotopes: Atoms of the same element that have the same number of protons and electrons but different numbers of neutrons.
metal: An element on the left of the periodic table. Metals are good conductors of heat and electricity.
molecule: A particle that contains atoms held together by chemical bonds.
neutron: A tiny particle with no electrical charge found in the nucleus of an atom.
nonmetal: An element at the right-hand side of the periodic table. Nonmetals are liquids or gases at normal temperatures. They are poor at conducting heat and electricity.
nucleus: The center of an atom. It contains protons and neutrons.
periodic table: A chart of all the chemical elements laid out in order of their atomic number.
pH: A scale that measures how acidic something is when it is dissolved in water.
products: The substances formed in a chemical reaction.
protium: The most common isotope of hydrogen, having no neutrons.
proton: A tiny particle with a positive charge. Protons are found inside the nucleus of an atom.
salt: The result of a reaction between an acid and a base.
tritium: A very rare isotope of hydrogen.

Index